ANAM MUHAMMAD

Engineering for Beginners

For a Student, by a Student

Copyright © 2023 by Anam Muhammad

All rights reserved. No part of this publication may be reproduced, stored or transmitted in any form or by any means, electronic, mechanical, photocopying, recording, scanning, or otherwise without written permission from the publisher. It is illegal to copy this book, post it to a website, or distribute it by any other means without permission.

Anam Muhammad asserts the moral right to be identified as the author of this work.

Anam Muhammad has no responsibility for the persistence or accuracy of URLs for external or third-party Internet Websites referred to in this publication and does not guarantee that any content on such Websites is, or will remain, accurate or appropriate.

Designations used by companies to distinguish their products are often claimed as trademarks. All brand names and product names used in this book and on its cover are trade names, service marks, trademarks and registered trademarks of their respective owners. The publishers and the book are not associated with any product or vendor mentioned in this book. None of the companies referenced within the book have endorsed the book.

First edition

This book was professionally typeset on Reedsy.
Find out more at reedsy.com

Engineering is a beast, but one that can be tamed.

Contents

1 Welcome!	1
About the Author	1
What to Expect	2
2 Introduction to Engineering	3
What is Engineering?	3
What do Engineers Do?	4
3 The History of Engineering	6
Civil and Military Engineering	6
Ancient Civilizations	7
4 Types of Engineering	10
Interdisciplinary Fields	10
5 Civil Engineering	12
Architectural Engineering	15
Structural and Construction Engineering	17
6 Mechanical Engineering	20
Industrial and Systems Engineering	23
Aeronautical Engineering	25
Aerospace Engineering	27
Mechatronics and Robotics	29
7 Electrical Engineering	31
Computer Engineering	34
Computer Science	36
Software Engineering	38
8 Chemical Engineering	40

Nuclear Engineering	43
Biochemical and Biological Engineering	45
Environmental Engineering	47
9 Interdisciplinary Fields of Engineering	50
Materials Science and Engineering	50
Biomedical Engineering	52
Bioengineering	55
10 Engineering Ethics	57
Code of Ethics	58
Case Study: The Space Shuttle Challenger Disaster	58
11 Preparing for an Engineering Career	63
Education	63
High School	64
College Classes	65
Computer-Aided Design	66
Software for Engineers	67
Career Paths	67
12 Conclusion	69
13 Resources	70

1

Welcome!

Hello, dear reader, and welcome to "Engineering for Beginners: For a Student, by a Student"! My name is Anam Muhammad, and I am beyond excited to be your guide into the world of engineering. No matter how vast or limited your knowledge in engineering is, I am extremely grateful that you decided to pick this book up and give it a chance.

About the Author

Who am I? I am a biomedical engineering student at the Georgia Institute of Technology in Atlanta, Georgia. Since my childhood, I've enjoyed building, asking questions, and learning details that many would not bother themselves with. I have a passion for innovating and efficiently improving the quality of human life. I've always known that I wanted to pursue technology and medicine and ended up discovering biomedical engineering to be my calling (see the subchapter "Biomedical Engineering" for more of my story).

What to Expect

The reality is that many students are unsure about the type of engineering to pursue - or if they even want to. "Engineering for Beginners: For a Student, by a Student" was conceived to provide insight, knowledge, and guidance for students interested in the vastness of engineering. This book does not teach the concepts mentioned, it solely conveys general information about engineering and its disciplines. This is especially valuable to those who want to educate themselves about the various types of engineering.

With that said, this book does not need to be read from cover to cover. The chapters are divided in a way that anyone can skim through the table of contents and find what they would like to read. The bulk of the chapters centers around the types of engineering where you'll learn about their context, subjects, skills, and careers. We'll even get to do an engineering case study to see how engineering ethics come into play in the real world.

With that said, feel free to explore! Never be afraid to let your mind wander.

2

Introduction to Engineering

What is Engineering?

What comes to your mind when you hear the word "engineering"? Perhaps you think of science and math, and you would be correct. But, that isn't all. There is a lot more to it than learning scientific concepts and equations. Engineering is the application of science and mathematics with the ambition of innovating or developing products for technological use. Engineers solve problems and once we do, we look for more!

"Engineering" is derived from the Latin word *ingenium*, which means "cleverness", and the word *ingeniare*, which means "to design". When you look at the world around you, these two words begin to make a lot of sense. There are countless examples, here are just a few:

- Computers

- Buildings
- Athletic wear
- Game consoles
- Cars
- This book

"This book"? Certainly! Specifically, *how* you are reading this book. Be it a print or electronic copy, people had to cleverly design the printing, distribution, and communication processes for you to be able to read these words.

The principles of engineering are based on physics, chemistry, and mathematics. The fundamental concepts and theories of these subjects originated thousands of years ago. They have been tested and refined since then. These ideas may be old, but they are very much still relevant. Engineering would not be the same without them.

What do Engineers Do?

Engineers use science and math as tools to solve problems in the world. They tackle issues and make improvements with resources, materials, and energy. Engineers design, analyze, and build.

It is also worth noting that it is not always necessary to invent new products or systems. Not all engineers are inventors because not all of us need to be. Engineers are also responsible for repairing, optimizing, and innovating. These actions are equally as critical as creating unique products. What's the point of constructing hundreds of machines if you can't keep any of them working? Truly, engineers value quality far more than quantity.

Hard skills are skills and abilities that are obtained through some form of education and/or practice. Examples of hard skills are coding, and speaking languages. Soft skills, on the other hand, are personality traits and social skills. An engineer's success depends on soft skills just as much as hard skills. Being open to feedback, constructive criticism, problem-solving, and critical thinking are valuable traits. Communication and teamwork are key since many projects require a hefty amount of collaboration. Engineers use their hard skills, soft skills, and knowledge to improve our lives.

3

The History of Engineering

We think of engineering as a modern subject that started with advanced technology like the Internet, cell phones, and computers. The truth is, engineering has been around for a significant amount of time. It was born ever since people began thinking creatively to solve problems.

Civil and Military Engineering

Let's go back in time to when ancient civilizations existed. Unlike today, conquest and protection motivated survival. Ancient empires had a protect-or-destroy mindset. Thus, the history of engineering has its roots in military engineering. Military engineering is the building of military weapons (catapults, assault towers, etc.), works, and methods of communication. Later, civil engineering emerged. Civil engineers were given this title to distinguish them from military engineers. They were involved with buildings, bridges, roads, and other infrastructure.

Ancient Civilizations

The earliest recorded engineer was an Egyptian named Imhotep. He lived during the reign of Pharaoh Djoser and was one of his trusted officials. It is believed that Imhotep was responsible for the designing and construction of the pyramid of Djoser around 2600 BCE in Saqqarah, Egypt. He may also have been the first person to use columns in architecture. From his work, we consider Imhotep to be a civil engineer.

The pyramid of Djoser. It is also called the Step Pyramid

Other ancient civilizations had remarkable achievements as well. The Incas and Aztecs of South America also built impressive pyramids that were completely independent of their African counterparts. In Rome and Persia, aqueducts brought a fresh and clean flow of water to people throughout the empires. Ancient Greece is home to the Parthenon, a temple dedicated to the Greek goddess Athena. China has an incredible history of engineering; the Great Wall of China was built under Qin Shi Huangdi during the Ming Dynasty and is the largest man-made architecture. Heading over to South Asia, we have the Indus River Valley (modern-day Pakistan and India), where the ancient villages of Harappa and Mohenjo-Daro resided. These two ancient villages had impressive irrigation systems long before the Romans.

Some of these monuments still stand as a testament to the skills of engineers during their time. Today, the Great Wall of China spans 5,000 miles and Harappa's ruins can still be visited. Others, while no longer erect, still represent the pinnacle of engineering of their time. Many of them are among the Seven Wonders of the Ancient World, such as the Great Wall of China. These structures truly never cease to amaze.

As mentioned earlier, these feats serve civil and military purposes because of the need for safety and basic comfort. Eventually, conquest vanished and country borders were established. Humans moving away from military and civil engineering paved the way for other engineering disciplines to thrive.

4

Types of Engineering

Engineering is a broad term that includes a myriad of disciplines. There are four main ones: Mechanical, Electrical, Civil, and Chemical Engineering. Within each of these disciplines are more subdisciplines. We won't go through all of them since there are so many fields of engineering, but we will talk about the most prevalent ones.

Of course, there are far more engineering disciplines beyond what is described here. If there is a field of engineering you can not locate here (which may very well be the case), you should use other resources. In fact, for anyone interested, I would highly recommend doing your own external research.

Interdisciplinary Fields

Although this book lists them separately, it is important to know that many engineering fields combine multiple disciplines. They are referred to as interdisciplinary fields. The categorizing of disciplines varies from source to source; you may find that other people sort them differently, and that is completely acceptable. Disciplines that I believe are very broad to organize under one of the main fields can be found in the chapter "Interdisciplinary Fields of Engineering". Enjoy the read!

5

Civil Engineering

Context

Civil engineering is one of the oldest disciplines of engineering. It dates back thousands of years ago. Humanity is - quite literally - built on the work of civil engineers. In the 18th century, an Englishman named John Smeaton proclaimed himself a civil engineer. He designed bridges, canals, harbors, and lighthouses, earning himself the title of the "father of civil engineering". Smeaton was also a mechanical engineer and physicist. His work was vital to the development of cement.

In regards to urban design, Hippodamus of Miletus is credited for geometric community planning. His ideas are why cities are built in a rectangular design and why streets are straight and intersect at right angles. This urban layout and division of cities are referred to as the Hippodamian Plan, and its concepts are still used today!

What do Civil Engineers Do?

CIVIL ENGINEERING

Civil engineers design and maintain the systems, technologies, and structures to support the human population. They invent technologies for challenges that have yet to come. From our short history lesson, we learned that civil engineering concerns buildings, bridges, roads, and other infrastructure. A civil engineer creates the blueprint and foundations for a functional society and environment.

ENGINEERING FOR BEGINNERS

Subjects and Skills:
- Computer-aided design (CAD)
- Structural design
- Hydraulics
- Surveying
- Modeling

Careers:

- Architecture
- Building industry
- Construction
- Government service
- Urban planning
- Environmental sustainability

Architectural Engineering

It should come as no surprise that architectural engineering falls under civil engineering. Architectural engineers use engineering principles for buildings and their systems. They are not to be confused with architects. Architects specialize in the construction and aesthetics of buildings while architectural engineers apply theoretical and practical knowledge to engineer systems within those buildings. They bridge the gap between architecture and engineering.

Subjects and Skills:

- Structural design
- System integration
- Hydraulics
- Modeling

Careers:

- Architecture
- Construction
- Government service
- Environmental sustainability

Structural and Construction Engineering

Structural engineering concentrates on the framework and design of structures while construction engineering gets more involved in the physical building. Together, structural and construction engineering support civilization with infrastructure, and facilities that make life comfortable.

A scaffolding structure is temporarily set in place to support workers during construction.

Subjects and Skills:

CIVIL ENGINEERING

- Computer-aided design (CAD)
- Structural design
- Hydraulics
- Modeling

Careers:

- Construction
- Government service
- Environmental sustainability

6

Mechanical Engineering

Context

Mechanical engineering is one of the most diverse engineering disciplines. Although a discipline on its own, mechanical engineering is further divided into specialized fields. Mechanical engineering has solved problems and inconveniences in all parts of our lives. We can even say that it began as early as the invention of the wheel because before motor vehicles existed, people either traveled by foot or by mounting an animal. The assembly of carriages added itself as a mode of transportation, and it couldn't have happened without the wheel!

Thomas Savery and Thomas Newcomen are credited for inventing the steam engine which initiated the Industrial Revolution. The Industrial Revolution was a significant turning point for mankind that occurred during the 18th and 19th centuries. Machines played a major role during this time. James Watt, a mechanical engineer, optimized the steam engine by condensing steam in a separate chamber rather than the piston chamber.

What do Mechanical Engineers Do?

Mechanical engineers develop and maintain machinery, mechanical systems and processes, and materials. Simply put, they study objects and systems in motion.

Subjects and Skills:

- Computer-aided design (CAD)
- Product design
- Machine design and maintenance
- Modeling

Careers:

- Manufacturing
- Automotive industry
- Aviation
- Industrial systems
- Mechatronics
- Robotics

Industrial and Systems Engineering

Industrial and systems engineers support other engineers by designing and optimizing the facilities that they use. The other engineering fields we learn about tend to concentrate on a product or an outcome, but these fields are extremely process-oriented and view systems as a whole first. Industrial and system engineers blend math, science, and business while simultaneously considering human factors (ie. workers and communication) to form efficient and safe processes, systems, and programs.

Subjects and Skills:

- Project management
- Psychology
- Management
- Optimization
- Constructive criticism

Careers:

- Industrial engineering
- Engineering Management
- Consulting

Aeronautical Engineering

Aeronautical engineers make aircrafts like planes, helicopters, and jets. The Wright brothers, Wilbur and Orville Wright, were contemporary aviation pioneers. They designed the first powered aircraft with a gas engine, which Wilbur Wright flew for 255 meters for 59 seconds, equating to about 9 miles per hour. Although not as fast as today's planes, the accomplishments of the Wright Brothers and World War I instigated the spark of aviation.

To gain the upper hand in combat, engineers innovated more efficient aircraft designs to handle military loads. Soon, humans began to reach high speeds and altitudes. Our mastery of the skies made way for commercial airlines. Aeronautical engineering was able to connect the world through another travel method that wasn't automotive or marine transportation.

A plane jet engine. The turbine consists of rotating blades. It is powered by high-pressure air from internal combustion.

Subjects and Skills:

- Computer-aided design (CAD)
- Aerodynamics
- Flight mechanics
- Propulsion and combustion
- Materials science
- Aircraft design
- Optimization

Careers:

- Airline industry
- Aircraft management

- Military and Defense

Aerospace Engineering

Aeronautical engineering is limited to aircraft, but aerospace engineering includes spacecraft design and research. Aerospace engineers build space shuttles, rovers, and satellites to address challenges from space, solar systems, and extraterrestrial bodies. Aerospace engineering made enormous advancements during the 20th century during the Space Race - a time during which the United States and Soviet Union vied for spaceflight dominance. This rivalry showed just about competition is a driving factor for innovation and improvement, which are qualities an engineer should always strive for.

There is still much to uncover in aerospace engineering and there are always space missions being planned. People went to the Moon, and it's possible some might live on Mars one day. If you're interested in aerospace engineering, take a look at the Challenger space shuttle case study in "Engineering Ethics"!

Subjects and Skills:

- Computer-aided design (CAD)
- Aerodynamics
- Flight mechanics
- Propulsion and combustion
- Materials science
- Spacecraft design
- Optimization

Careers:

- Aeronautics
- Aerospace
- Space exploration
- Military and Defense

MECHANICAL ENGINEERING

Mechatronics and Robotics

For most of history, humans have been the main unit in the workforce. But we have our limits. We get tired and have limited strength and mobility. To increase our efficiency, we started building and employing robots and automated systems. These have allowed us to manage our time and energy effectively while also minimizing safety risks.

Mechatronics is a combination of mechanical, electrical, and computer engineering. Robotics is a branch of mechatronics that specializes in automated devices and bodies. There are multiple pathways you could take by pursuing mechatronics or robotics. For example, you can study medical robotics and biomechanics, industrial automation, or artificial intelligence. The first industrial robot, Unimate, was developed by George Devol Jr. and Joseph Engelberger around 1960. The Unimate robot had six programmable axes of motion that could lift up to 225 kilograms (about 500 pounds) at relatively high speeds.

Automatic industrial robots

Subjects and Skills:

- Computer-aided design (CAD)
- Electronics and circuits
- Sensors

Careers:

- Industrial engineering
- Automotive industry
- Intelligent systems
- Medical robotics
- Artificial engineering

7

Electrical Engineering

Context

Electrical engineering applies electricity and electromagnetism to electronic devices and systems. Its foundations came to light in the 1800s when electricity was introduced. Once electrical power was discovered, engineers realized that they could harness it for machines. Electricity functions as the "blood" of electronics, a field born in the late half of the 19th century. Electrical development was accelerated by the invention of the transistor in 1947.

The experiments of multiple people and their knowledge affect what we have today: the Internet, electric lights, robots, and more! For example, there was Alessandro Volta, Michael Faraday, and Georg Ohm. If you have ever heard of voltage, Faraday's law, and the ohm, you can thank these people! Similar to how the steam engine was an industrial revolution, their work paired with the telegraph was revolutionary to electrical engineering.

What do Electrical Engineers Do?

Electrical engineers analyze and design electrical equipment, systems and controls, and processing. They understand how to integrate electricity and circuits into a variety of fields like mechanics, optics, energy, electromagnetics, medicine, and telecommunication. The results are power generation and automated control systems!

ELECTRICAL ENGINEERING

The process shown is called soldering. A solder is a metal alloy used to fuse wires and metal. Solders have a low melting temperature that gives them this property.

Subjects and Skills:

- Circuits
- Electronics
- Electromagnetics
- Sensors and transducers

Careers:

- Control Systems
- Electrical instrumentation
- Network engineering

Computer Engineering

The rise of computer technology in the 20th century emphasized the need for computer design, computer networks, and computer applications. It's become so integrated into our lives that it's difficult to imagine living without them! Therefore, the role of computer engineers is vital since they are the ones who create and improve computers. They implement hardware components - the physical parts of a computer like circuit boards, processors, memory devices, and routers.

Computer engineers are expected to master the connection between mechanical and electrical engineering. Almost all computers have two core parts: a memory that stores information and a central processing unit (CPU) that executes programs. Computer engineers build and

ELECTRICAL ENGINEERING

produce the computer systems that are used today. We thank them for making effective computers, which in turn, makes for generations of effective engineers!

A circuit board. Circuit boards electrically connect components in a device so that an electric signal can be processed.

Subjects and Skills:

- Computer-Aided Design (CAD)
- Circuits
- Electronics
- Sensors

Careers:

- Mechatronics
- Computer architect
- Machine learning
- Software development
- Hardware development
- Computer security

Computer Science

Let's talk about computer science! It's not a field of engineering like computer [hardware] engineering because it primarily focuses on software and computer theories. Instead of building computers, computer science builds performance. Studying this field opens a plethora of opportunities and courses that you can specialize in. Here is a handful of them:

- Videogame design
- Robotics
- Data analysis
- Website design and maintenance

Computer science is a needed and ever-growing technological area. It requires less hands-on interaction and is more theory-based than computer engineering.

ELECTRICAL ENGINEERING

Subjects and Skills:

- Coding
- Programming
- Debugging

Careers:

- Game development
- Artificial intelligence
- Ethical hacking
- Software development
- Web design

Software Engineering

Unlike computer engineers who improve hardware parts, software engineers create computer programs, or software. They maintain them as necessary to achieve varying results with the hardware a computer engineer provides. The chapter "Preparing for an Engineering Career" has a subchapter titled "Software for Engineers" that mentions specific software engineers use.

I'm sure most of us are familiar with apps on our phones and devices. Those are examples of software applications - they aren't tangible hardware. Rather, they are developed using software engineering skills. Software engineers also develop and fix operating systems, applications, and programs to solve real-world problems. When your phone and apps release new updates and bug fixes, there is a high chance a software engineer was behind it.

ELECTRICAL ENGINEERING

Subjects and Skills:

- Coding
- Programming
- Debugging

Careers:

- Programming
- Software development
- Game design
- Web design

8

Chemical Engineering

Context

Chemical engineering developed in the late 19th century because the Industrial Revolution introduced a supply and demand of new materials and processes. The production of chemicals was highly necessary for large-scale manufacturing. However, it is necessary to break down chemical processes into *unit operations*, which are individual steps in the production phase. Unit operations are founded on fundamental science principles like momentum, mass conservation, and thermodynamics.

What do Chemical Engineers Do?

Chemical engineers design chemical plants and processes that convert chemicals, raw materials, and energy for production. They are experts in chemical reactions that occur in metallurgy, textiles, medicine, and the environment. Many chemical engineers today are employed in the food industry and are responsible for artificial flavoring.

Chemical reactions are the basis for chemical engineering applications. While they can be conceptualized, unfortunately, not all reactions take place during a practical amount of time. Some chemical reactions can take years to occur because their *activation energy*, or energy needed to initiate a reaction, is too high. This is where reaction kinetics comes into play. Kinetics seeks to understand the rate of chemical reactions. Fields where chemical engineering applies extend across medicine, solar energy, thermodynamics, biofuels, and air quality control.

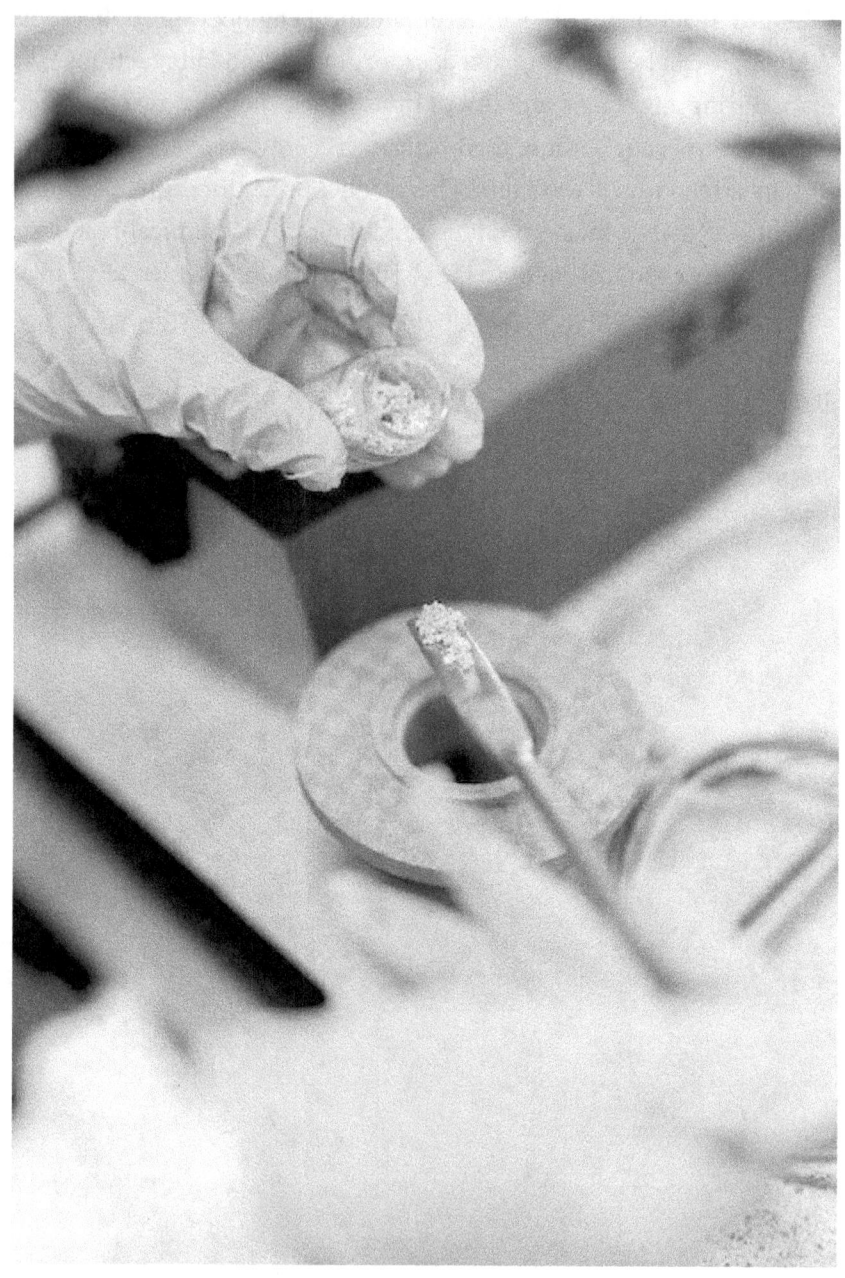

Subjects and Skills:

- Catalysis
- Reaction kinetics
- Optimization

Careers:

- Government
- Chemical processing
- Healthcare
- Pharmaceutics
- Drug delivery
- Food engineering
- Sustainable energy

Nuclear Engineering

Nuclear engineers harness energy from nuclear reactions. They study and research nuclear reactor cores, systems, and operations. Beyond that, there are nuclear power economics, radiation sources and detection instruments, radiation protection, and regulatory requirements.

Nuclear engineering is currently a struggling discipline. After the United States dropped two atomic bombs on Japan (one on Hiroshima and the other on Nagasaki), nuclear engineering repelled many people due to a developed social and environmental stigma. This lack of interest led to an entire missing generation of nuclear engineers. To combat the declining student enrollment in nuclear engineering fields, organizations and scholarships are attempting to incentivize students.

A nuclear power plant (NPP) utilizes heat released from a nuclear reactor to vaporize water into steam. Steam propels turbine generators to generate electricity. In America, boiling water or pressurized water reactors are used.

Subjects and Skills:

- Reaction kinetics
- Optimization
- Radiation safety protocols

Careers:

- Military and defense
- Weapons
- Sustainable energy
- Environmental health
- Food irradiation

Biochemical and Biological Engineering

Biochemical and biological engineering combine life and physical sciences for medical, pharmaceutical, dental, veterinary, and physical therapy applications. They explore chemicals and biological matter like cells, enzymes, and antibodies to create new products.

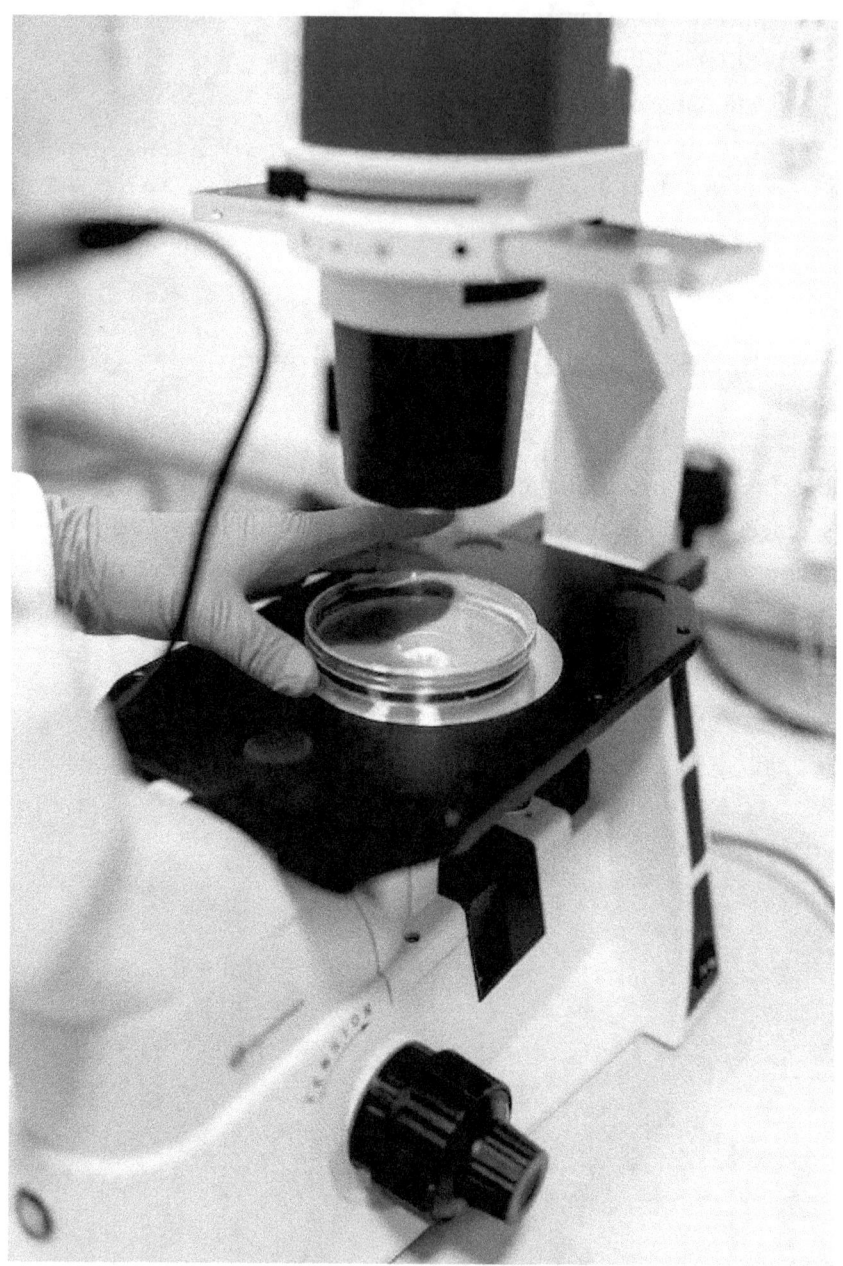

A Petri dish housing microorganisms is examined microscopically.

Subjects and Skills:

- Bioprocessing
- Research
- Cell biomanufacturing
- Microbiology

Careers:

- Research and Development
- Pharmaceutics
- Food engineering
- Biomedicine
- Quality control

Environmental Engineering

Environmental engineering uses engineering practices and science to address environmental issues such as sustainable air, water, and land resources. Environmental engineers target improvements in human health and environmental restoration. The world is our home, and environmental engineers make sure that we adapt well to it while also taking care of it. They tackle matters related to energy transfer, water sourcing, waste disposal, environmental degradation, and food production.

Living in countries with drastically different resources and geography, environmental engineers take advantage of *globalization,* the process of human and resource dispersion. Globalization has allowed engineers to collaborate and share resources and ideas with people of all nationalities. Environmental engineering is a constantly growing field that adapts to the world around us.

One challenge that environmental engineering is trying to overcome is the use of renewable energy. The processes that engineers implement in product development rely heavily on non-renewable energy. Non-renewable energy comes from non-renewable resources, which are limited in availability.

This means that humans use non-renewable resources like fossil fuels at a faster rate than they naturally form. Another dilemma is that waste emissions like carbon dioxide are unhealthy for the environment. Environmental engineers seek to address these issues by researching sustainable resources in the form of renewable energy.

CHEMICAL ENGINEERING

Solar and wind energy are renewable resources

Subjects and Skills:

- Energy analysis
- Urban systems
- Economics
- Research

Careers:

- Consulting
- Research
- Air quality control
- Environmental health

9

Interdisciplinary Fields of Engineering

Materials Science and Engineering

We use materials to make objects. Ever since the Bronze Age, humans used elements to forge materials like - you guessed it - bronze! The notion of finite and tangible materials is what materials science and engineering revolves around. Think of it as a mix of chemistry and mechanical engineering.

Materials Science is a vast discipline since there are so many materials available to us: new, old, and even ones yet to be made in the future. Materials science engineers perform tests to determine material properties. To achieve desired properties, they process materials to achieve certain structures. The flow of process, structure, and properties is the basis of materials science and engineering. Materials are classified into the following groups: polymers, ceramics, metals, and composites.

The lens shown magnifies the content on the paper. Materials science engineers would study the microstructure of the lens to understand its optical properties.

Subjects and Skills:

- Research and Development
- Chemistry
- Physics
- Biology
- Optimization

Careers:

- Industrial processing
- Research
- Nanotechnology

- Biomedicine
- Drug delivery

Biomedical Engineering

Earlier, I promised to share my engineering journey. Ever since my childhood, I have been interested in medicine and technology. As I went through middle and high school, I took healthcare science and advanced math classes. In high school, I discovered biomedical engineering and how it applied both of my passions to enhance and save lives. I was sold! I started my biomedical engineering degree and never once looked back.

Biomedical engineering is a relatively new field of engineering that officially emerged in the 1960s. Biomedical engineers design and build medical devices, processes, and equipment to solve clinical problems. Engineering skills and principles are applied to science and medicine. Biomedical engineering seeks to advance healthcare; from the analysis of medical conditions to diagnosis, to treatment and recovery. Our ultimate goal is to improve the quality of human life.

Due to our direct interaction with the human body, we must ensure that we minimize medical repercussions as much as possible. This is critical for fields like tissue engineering and biomaterials since our bodies react to foreign materials like metals and ceramics. To avoid these problems, biomedical engineers need to know about mechanical and electrical engineering, materials science, anatomy, biology, and chemistry. It is an extensive discipline but without it, artificial organs, defibrillators, pacemakers, and other medical technology would not exist today.

ENGINEERING FOR BEGINNERS

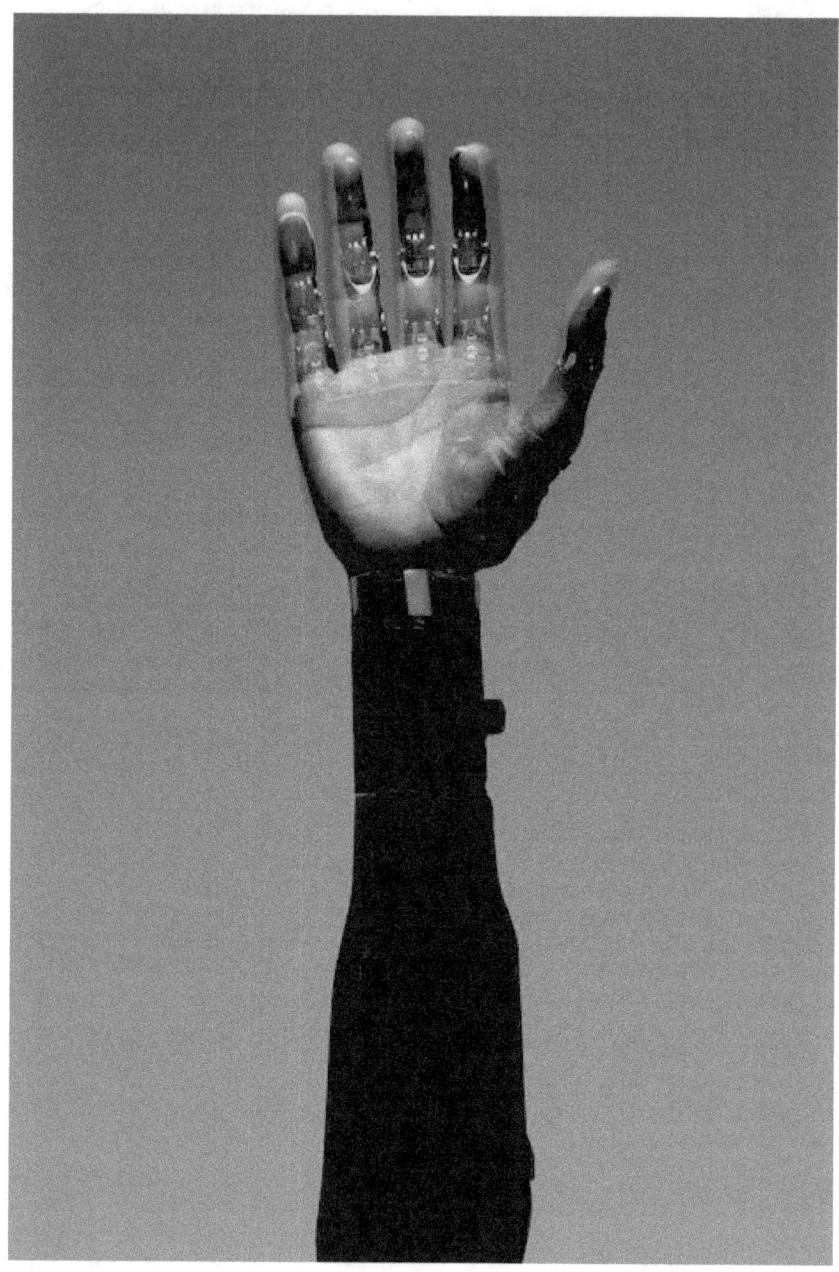

INTERDISCIPLINARY FIELDS OF ENGINEERING

Subjects and Skills:

- Computer-Aided Design (CAD)
- Medical device design
- Biomechanics
- Biostatistics and data analysis
- Electronics and circuits
- Materials science

Careers:

- Healthcare
- Research
- Medical imaging
- Bionics and prosthetics
- Tissue engineering
- Drug delivery

Bioengineering

Bioengineering may sound as though it has an identical role to biomedical engineering, but that is not the case. Biomedical engineering concentrates on human and animal biology for medical applications while bioengineering works with other biological systems, like plants and microorganisms.

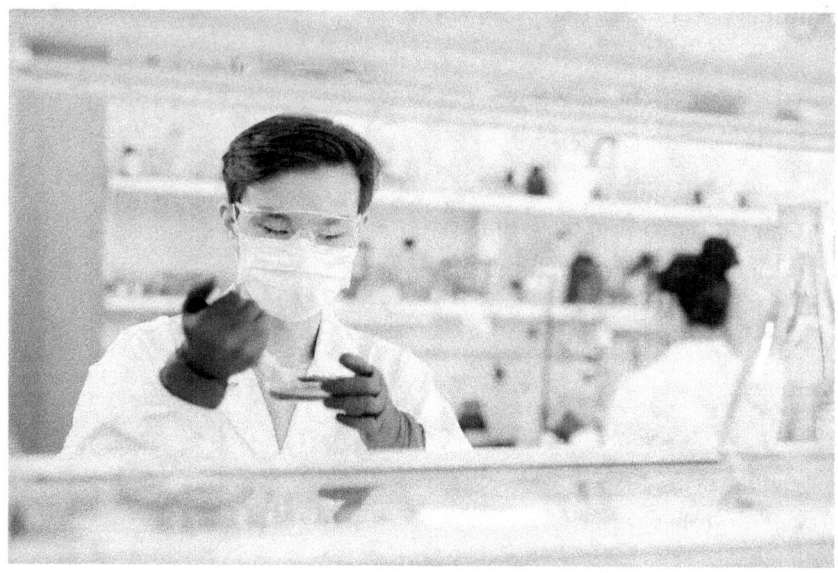

Subjects and Skills:

- Bioprocessing
- Cell biomanufacturing
- Microbiology
- Research
- Materials science
- Biostatistics and data analysis

Careers:

- Research and Development
- Genetic engineering
- Biomedicine
- Biomaterials
- Food engineering

10

Engineering Ethics

I like to talk about the moral aspects of engineering because I feel as though it isn't talked about enough, at least in the American education system. When students get introduced to engineering, they hear positively connotated terms like "creating", "solving", and "inventing". But what about if I threw the word "failure"?

Failure is nothing to be afraid of. Everyone encounters it in their lives - not just engineers. Failure can strike our self-esteem and confidence down. Engineers run into obstacles all the time, but when our proposed solutions don't work as intended or even fail, the degree of the consequences can range from a simple nuisance to a devastating tragedy.

Ethics is a moral philosophy that tries to separate wrong from right, good from bad. In the same way, engineering ethics keep engineers in line with a sense of morality to ensure that we are doing our work to the best of our ability. As a result, our outcomes are representative of quality work and practice.

Code of Ethics

Engineers follow something called a *Code of Ethics*. There are different kinds, but the core values are essentially the same.

1. **Safety:** Safety is the #1 priority for all engineers. Designs and ideas should only be accepted if they conform to established engineering standards.
2. **Competence:** Engineers should only work in areas they are knowledgeable about.
3. **Honesty:** Don't lie. If bad results don't show up now, they will later.
4. **Merit:** An engineer's reputation should culminate from their work and contribution and not by dishonest means.
5. **Respect:** Treat everyone with respect and dignity. Serves as general life advice, no?

Case Study: The Space Shuttle Challenger Disaster

On January 28, 1986, the space shuttle Challenger launched from the Kennedy Space Center in Florida. 73 seconds into its flight, Challenger exploded and broke apart. All seven crew members died. It was revealed that a leak in one of the solid rocket boosters let escaping gases ignite the main tank with liquid fuel, causing a cataclysmic explosion.

ENGINEERING ETHICS

The explosion of the space shuttle Challenger

How did this leak happen? Well, NASA discovered that the O-ring seal in the rocket booster failed. I will point out that this is only the tip of the iceberg. There was so much more to the story but at its core, the Challenger Disaster was the result of a flaw in engineering ethics.

O-Rings

O-rings are torus (donut-shaped) seals placed between parts during assembly. They need a sufficient gap and fluid pressure to seal a joint. Challenger's O-rings had a gap of 0.310 inches; however, standard O-

59

ring handbooks indicated that the grooves should have been between 0.375 to 0.380 inches. The result? The O-rings were too small to seal properly, hence the gaseous leak.

Furthermore, O-rings are thermoset polymers (a nod to materials science), meaning that they become brittle and break in cold environments. So when NASA chose to launch Challenger under cold temperatures, it was only a matter of time before the O-rings failed.

Engineering Ethics Gone Wrong

It's not as though the O-rings alone were to blame. Complications didn't arise from mechanically engineering the O-rings but rather from how these complications were addressed. A company named Morton-Thiokol constructed Challenger's solid rocket boosters. As Thiokol engineers were testing the boosters, they grew worried about their performance. The day before Challenger's launch, a teleconference between Morton-Thiokol and NASA took place, and the Thiokol engineers expressed their concern. Their official recommendation was not to launch Challenger.

A vote was conducted, but only among senior managers - none of the Thiokol engineers were included. The managers concluded that the engineers' data was inconclusive and that the frigid temperature posed no launch risks. NASA was also politically pressured by Congress since Challenger's launch had already been postponed before. So, despite the warning from the Thiokol engineers, the NASA managers proceeded with the launch of the space shuttle. Their decisions and lack of concern for the safety of the seven passengers of Challenger caused the spacecraft's traumatic catastrophe.

Lessons Learned

Losing lives is not a matter to be taken lightly. The Challenger disaster is a harsh reality to face but we must come to terms with it. Engineers need to learn from their mistakes and understand how their actions can violate the Code of Ethics. The Challenger case study emphasizes the role of components, the value of established data handbooks, and the significance of manufacturing variation tolerances.

NASA's internal structure also underwent major changes. Before

Challenger, NASA's safety culture was to not launch only if it was proved to be unsafe. Now, safety is no longer assumed. Aircrafts and spacecrafts are only launched if it is proven safe to do so. Although this new concept is a double-negative of the previous one, it is a better adaptation of an engineer's work.

11

Preparing for an Engineering Career

If you're intrigued to the point of pursuing engineering, congratulations! It is very rewarding. Here you will find general information and tips for navigating education in engineering in the United States. Everyone's journey is different; there is no right way to approach an engineering career. Some people switch to a discipline later in their lives and others already know which one they want to do.

Education

Formal engineering education in the United States begins in a university, but engineering can be introduced even earlier. Thanks to STEM programs, children and students in the K-12 curriculum can learn about science, technology, engineering, and math!

Many students who have an idea about which career they want to do tend to have an advantage when it comes to college applications. Some middle and high schools offer engineering classes that can help you discover what your interests are.

High School

STEM classes like chemistry, physics, biology, and math are already required and incorporated into the education system. However, I have found that math levels vary quite a bit. If your school offers them, I highly recommend taking AP (Advanced Placement) and/or IB (International Baccalaureate) courses by the College Board and International Baccalaureate, respectively. They offer college-level courses to high school students. After passing a cumulative exam, AP and IB students can earn college credit, boost their GPA (Grade Point Average), and save money on credit hours they would have taken in college anyway.

A quick note about AP and IB courses: you make the most out of the exams if you research the universities you plan to apply to. Not all schools accept certain AP/IB classes, and earning college credit depends on your exam score. Each university has a set minimum score they will accept. Not to worry if you don't pass the exams, though; you can choose to not report your scores to universities. They do not negatively impact you.

The college application season can be one of the most stressful moments in a student's life. Applications are open to senior high schoolers, but the earlier a student determines their field(s) of interest, the better. Here are some tips I have for preparing for college applications:

- **Be proactive:** Research careers and schools as early as possible.

Use your time wisely.
- **Initiate passion projects based on your interests:** This is my golden advice to all prospective students. Starting or joining a project about a topic you care about shows universities a high level of dedication. Bonus points if your project is relevant to your field of interest, but it doesn't need to be.
- **Develop skills with online resources and courses:** There are free resources virtually everywhere. Self-learning is a skill that needs commitment, so schools love seeing this
- **Connect with people:** Building personal and professional connections. Networking will only contribute as you move forward in your life; you never know how a person can help you, or vice versa.

College Classes

In university, an engineer's classes will differ from discipline to discipline. You will need to take specific classes based on your major. Engineering majors take physics and introductory chemistry courses at the minimum. Biology is added for suitable disciplines.

Mathematics is a longer story. Most engineers need to take upper-level math classes by the time they graduate. AP and IB credits become handy here because they let you "skip" corresponding classes. In ascending order, the order of math looks something like this:

1. College algebra
2. Precalculus
3. Differential Calculus (Calculus I)

4. Integral Calculus (Calculus II)
5. Multivariable Calculus (Calculus III)
6. Linear Algebra
7. Ordinary Differential Equations

As I mentioned, other classes are directly related to the type of engineer you want to be.

Computer-Aided Design

Most engineers will be familiar with some form of Computer-Aided Design, called CAD for short. This was listed earlier as a skill for many engineering disciplines. CAD is the process of using software that designs 2D or 3D objects on a computer. Designing precise and replicable components is key for engineers because having the ability to prototype an object before manufacturing saves time, money, and materials. With CAD software, you can even run simulations and test how a component would behave under various circumstances. Once a prototype is designed, engineers 3D print physical copies of the object for use!

Examples of CAD software include:

- SolidWorks
- Solid Edge
- AutoCAD
- Autodesk Fusion 360
- MechDesign

- Rhino

Software for Engineers

CAD lets engineers develop physical parts, but it's also important for engineers to code in other software to create programs, code, analyze, and compute complex calculations. Of course, there is more beyond what is listed, these are just the ones I've seen used the most:

- Microsoft Excel
- MATLAB
- JavaScript
- Java
- Python
- SQL
- C++

Career Paths

With a degree in engineering, you have an abundance of career options to choose from. From my connections and learning about other engineers, I've found most engineers fall into one of these categories:

- Graduate School
- Professional Engineer

- Engineering Management
- Engineering Consulting
- Business
- Medical School
- Law School

12

Conclusion

Hopefully, this book was beneficial and provided general but useful information about engineering. I truly enjoyed writing "Engineering for Beginners: For a Student, by a Student" and sharing my knowledge! Perhaps you now have a slight idea if engineering is right for you. Engineering is a beast, but one that can be tamed. It requires a lot of time and dedication, but if you enjoy it, you will find it to be very much worth it.

Best of luck with your pursuits! If you found this book contributive, I would greatly appreciate it if you left a favorable review on Amazon!

13

Resources

A Brief History of Engineering. (n.d.). https://www.streetdirectory.com/travel_guide/192894/careers_and_job_hunting/a_brief_history_of_engineering.html

BS Biological Engineering | UGA College of Engineering. (n.d.). https://www.engr.uga.edu/bs-biological-engineering

CrashCourse Engineering. (2018, May). [Video]. YouTube. Retrieved December 28, 2022, from https://www.youtube.com/playlist?list=PL8dPuuaLjXtO4A_tL6DLZRotxEb114cMR

Kirkey, J. (n.d.). *What is Engineering? Definition, introduction and a brief history – Engineering and Technology in Society – Canada.* Pressbooks. https://pressbooks.bccampus.ca/engineeringinsociety/chapter/chapter-1/

Mechatronics and Robotics Engineering – Queen's Engineering and Applied Science. (n.d.). https://engineering.queensu.ca/programs/undergraduat

RESOURCES

e/mre/

Nuclear power plants - types of reactors - U.S. Energy Information Administration (EIA). (n.d.). https://www.eia.gov/energyexplained/nuclear/nuclear-power-plants-types-of-reactors.php

Smith, R. J. (1999, July 26). *Engineering | Definition, History, Functions, & Facts.* Encyclopedia Britannica. https://www.britannica.com/technology/engineering

The Editors of Encyclopaedia Britannica. (2009, January 14). *Challenger disaster | Summary, Date, Cause, & Facts.* Encyclopedia Britannica. https://www.britannica.com/event/Challenger-disaster

www.ingramcontent.com/pod-product-compliance
Lightning Source LLC
Chambersburg PA
CBHW071145240526
45465CB00024BA/1785